William Nemos

Artificial Memory

The Grand Method of Making a Bad Memory Good, and a Good Memory

Better

William Nemos

Artificial Memory
The Grand Method of Making a Bad Memory Good, and a Good Memory Better

ISBN/EAN: 9783337412906

Printed in Europe, USA, Canada, Australia, Japan

Cover: Foto ©berggeist007 / pixelio.de

More available books at **www.hansebooks.com**

ARTIFICIAL MEMORY:

THE GRAND METHOD

OF MAKING

A BAD MEMORY GOOD,

AND

A GOOD MEMORY BETTER.

By WM. NEMOS,

PROFESSOR OF ARTIFICIAL MEMORY.

———•●•———

SAN FRANCISCO:

Published for the Author by
A. L. BANCROFT & COMPANY.
1873.

PRICE, ONE DOLLAR.

PREFACE.

The object of this little book is to make a bad memory good, and a good memory better, by teaching a system which will enable any one to use the natural memory to wonderful advantage, as well as to perform the extraordinary feats of memory with which Mnemonists astonish the world.

The system is based upon sound philosophical principles, and is so simple, that the youthful schoolboy can master it with the same ease as the accomplished scholar. It will be found of immense service to everybody, be it the business man, the farmer, the student or the lady. In short, it is useful to *anybody* in remembering *anything*.

It has been tested in public and in private, and the author is satisfied, in common with his pupils, that it is superior in simplicity and completeness to other systems, possessing all their advantages with fewer defects. By its means, the weakest memory is enabled to accomplish what would be beyond the power of the strongest memory, unassisted.

Scholastic authorities recommend it as strengthening to the intellect generally; while it diminishes the probability of over-taxing the faculties, removes anxiety, and, by greatly saving time, gives opportunity for further studies, or for enjoyments.

The first part of the book contains a select compilation from standard works on memory, showing how to improve this faculty in the ordinary manner, how to study, etc., besides giving an interesting account of the nature of memory, and of its relation to the other qualities of the mind.

The object of education is two-fold—to store the mind, and to train it. Artificial Memory aids both. No one can question its power of storing the mind with facts, who has witnessed the feats of memory achieved by mere boys after a short instruction. No one, who has himself acquired the art, can doubt its power to train the mind, to educate it, in the best sense of the term. It substitutes thought and imagination for mere mechanical and hurtful repetition; it transforms a painful task into a positive pleasure.

Some of the feats placed within the reach of everybody who uses this system are: To remember hundreds of words, facts and figures, after hearing or reading them over *once*, so as to retain, with astonishing accuracy, the details of a lecture, conversation, book, newspaper, etc. To commit to memory, with ease, long and difficult tasks, such as statistics, populations, heights of mountains, astronomical magnitudes, logarithms, etc.—history, geography—a boy may learn more geography in one hour by this means, than he would learn by months of ordinary study. To recollect, with certainty, whenever desired, any number of ideas, engagements, errands, etc., that may occur to the mind at any moment.

CONTENTS.

CHAPTER I.

MIND AND MEMORY.

It does not pay to have a bad memory. To be ten times as long as need be in mastering a given subject, is bad enough, but it is still worse to have to suffer from daily, almost incessant forgetfulness, the omission of *little* things, as they are called, which often result in great vexations, disappointments, and disasters. Frequent forgetfulness renders many people absolutely miserable. Half their thoughts are regrets that they did not say something, or do something at the right time. But it too often happens that no regrets, however bitter, and no efforts, however energetic, can afford another opportunity for securing the same results which might have been obtained with ease, had memory done her duty. One case of forgetfulness may cost a fortune, may nullify the unceasing efforts of years, may blight prospects for life.

We must banish this, the greatest enemy of the mind, by using the means before us of improving our memory; thereby securing not only self-reliance and exemption from harassing

2

anxiety, but, what is of higher importance, in-
creased vigor for every faculty of the mind.

Everybody admits that it is a great advantage
to have a good memory; but few have a clear
idea of what memory really is, and of the important role it plays in every act of the intellect.
Analysis of the mind will show that it is almost
all memory. Memory is generally said to be
"the retention of things perceived," consequently perception is supposed to be of primary
importance—the cause, of which memory is but
the effect.

But we shall soon see that readiness and accuracy of perception depends upon memory.
For instance, no sensible man would invite a
draper to select diamonds for him, nor a jeweler
to choose cloth, when he could obtain the jeweler's advice upon the diamonds, and the draper's
aid in buying cloth, and why? Because everybody knows that, although the draper might
look at the diamonds, he would not be able to
perceive their merits or defects, and so with
the jeweler in regard to cloth. The question
now arises, why two men, with equally good
eyes, should not see equally well? Because the
jeweler, having in his memory a great many
facts and circumstances connected with diamonds, might discover many beauties and de-

fects in them, that would escape the notice of the draper. The draper, again, with his larger remembrance of the qualities of cloth, would be more competent in his own particular branch. This difference in the memory of the two men, is the cause of the difference in their capability of perception. Readiness and power of perception, then, are determined by memory; and, it must further be borne in mind, that that which is perceived, is valueless, unless it is retained by the memory. Perception not only depends upon memory, but that which is often called perception, is in reality but memory. If you say, "I perceive the cars coming; they have just crossed the bridge," it is equal to saying: "1 know, by means of my memory, that the cars were at the bridge, but now they are nearer to me."

Careful reflection upon this subject will show that the minutest mental acts are, for the most part, memory. Many words arouse in the mind the influence of complex memory, although they are generally supposed to produce simple perception. Thus, the word "windmill" first suggests to the memory the idea *wind*, which, as we cannot see it, makes us think of something which we can see, associated in our memory with *wind*, and then we think of *mill*—some *mill* already in our memory gives us the idea. And mark, while pronouncing the second sylla-ble of the word "windmill," we must have in memory the fact that the word *wind* preceded it; otherwise we might think of a *watermill*, or any other *mill*.

Conception, imagination and invention, and all the creative powers of the mind, have their origin in memory. People have the idea that these faculties are self-creative, self-developing and independent of others; but every new idea springs, directly or indirectly, from an old one, although it is common for the ordinary thinker to be unaware of the origin of thought upon many subjects. Many seem to regard it as a fanciful theory, that for every mental effect there is a cause; but it is so, although the cause may not be apparent at all times. We are indebted to the genius of Milton for "Paradise Lost;" but the genius of Milton consisted of a vivid remembrance of Bible records, of expressive words, and of metre. Had he forgotten either, his genius would have been marred. Shaks-peare, "the immortal bard," gave us some of the finest specimens of imagination that have ever charmed the world; but his genius con-sisted in an accurate remembrance of historical facts, of feelings and sensations of the human heart, and of the laws of rhetoric. Had he forgotten the characteristics of the individuals of whom he wrote; had he forgotten the objects and circumstances by which they were surrounded; or had he forgotten the plots which his brain had interwoven, the soul-inspiring name of Shakspeare would be a name unknown. We are indebted to Morse for the telegraph; but the genius of Morse consisted in the remembrance of a variety of objects, facts, principles and requirements, prompted by which remembrances, he tried experiments, which he remembered, until, by the perfection of the association of remembrances, the telegraph appeared to an admiring world.

Judgment — lofty, reverenced judgment — is humiliatingly dependent upon memory. Memory may exist without judgment, but judgment cannot exist without memory. The judge reviews the forensic knowledge gained from books and experience, which has been stored in memory during years of labor, and is enabled to render his verdict.

Shrewdness, or tact, depends mainly upon memory. The sharp business man is he who remembers men and manners.

Arithmetical calculation is the remembrance of certain numerical facts, rules and results.

Disposition depends greatly upon memory. If a youth is clever, but unreliable; if a man is talented but vicious, it is because certain moral remembrances are faint, or overshadowed by others. The memory of the heart, of the soul, requires exercise and development.

Love, delight, pleasure, arise from the remembrance of that which is in harmony with our condition.

Gratitude, appreciation, are the remembrances of benefits and advantages received.

Hope is the desire of the fulfillment of a remembered imagination. Thus, suppose an invalid seeks change of air in the hope of thereby becoming convalescent; he imagines that it may bring about his recovery, as it has reëstablished the health of others, and all this he remembers and desires.

Charity arises from the remembrance of facts and thoughts which awaken our sympathy. Want of charity is commonly called "forgetting to make allowances" for circumstances, or for the frailty of humanity.

Humility is the remembrance of our faults and weakness, our dependent condition, and our imperfections.

Prudence is the offspring of memory. The remembrance of past circumstances of a painful nature makes us anxious to avoid their recurrence. "A burnt child dreads the fire," that is, he remembers that there is or may be danger.

Honor is based upon the remembrance of duty. Many dishonorable acts result through forgetfulness, which may be bitterly regretted.

Conscience is but memory. It is the remembrance of what is supposed to be right. I say, "of what is *supposed* to be right;" because it is quite possible for us to be conscientious, and yet absolutely wrong and wicked. Conscience does not trouble the cannibal while feasting on his fellow-creature's remains, because no impression of the sinfulness of the act has been made upon his memory; and even if he should experience some gentle reminder, it would only be the result of his recollections, from own minor experience, that the sufferings of his victim were painful, as they would be to himself in case of a reprisal. Conscience is the battle-field selected by inclination and duty, on which to settle their strife. Where no duty is laid down by law, inclination follows her own dictates.

We might thus proceed with every mental manifestation, but the illustrations given are sufficient to show that memory, instead of being a comparatively mean quality of the mind—one which may be slighted—is the *basis* of intelligence; the faculty upon which the other operations of the mind wholly rest and depend. That it is not only the basis of the intellectual edifice, but the chief material of its body, and of its summit, without which the other faculties could not exist, and with which they are so inseparably connected, that whatever tends to dim the memory, must tend to diminish their lustre also. It follows, then, that the development, culture and preservation of the memory are matters of the very first importance.

As the brain is the recognized organ of the mind, and as its size, quality and activity are supposed to determine our mental power, it is thought that those only can have a good memory who have a certain kind of brain. That organization is necessary for the manifestation of mind and memory, is unquestionable; but to suppose that memory is wholly dependent upon it, is a grave error; one that may often prevent the proper effort to obtain improvement. Whatever physical cause tends to strengthen the brain, must certainly be advantageous to the mind and memory, and whatever is detrimental to the brain, is injurious to the mind and memory also. To obey the physical laws of our being is thus of primary importance, if we would have memory in perfection. But organization and health alone will not determine our powers of memory; there are two other essentials—*exercise* and *system*—and whoever possesses sufficient endowment of brain for the manifestation of *ordinary intelligence*, has enough brain to ensure, by these means, what is termed a *very good* memory. Inactivity destroys the memory; but though much may sometimes be accomplished in the way of remembrance by working *hard*, still more can be done by working *well*, and this brings us to system.

HOW TO IMPROVE THE MEMORY.

OBSERVE, REFLECT, LINK THOUGHT WITH THOUGHT, AND THINK OF THE IMPRESSION.

This is one of the best and most comprehensive rules ever given for the improvement of the memory. The following hints will illustrate some of its diversified applications:

COMMITTING TO MEMORY.

When you wish to learn a piece of prose or verse, try to grasp its general meaning first, and then particularize, that is, observe what words are used, and how they are placed.

Learn one sentence thoroughly by reflection, before you attempt to master another, and link them together by noticing how they follow. When you think you have succeeded in getting a sentence to run upon your tongue correctly, think of the impression; remove your eyes from the paper, and articulate the words aloud or mentally. Immediately afterwards cover the sentences with your hand, and again repeat, allowing yourself to look for each word just after you have uttered it. You will then detect any error of omission, substitution or transposition. Many people recommend writing out, a great many times, that which you would learn; but this is not so good as the plan just suggested.

If you have learned anything by ear, and are fearful of forgetting it, write it out once clearly, and afterwards look at it carefully; this will give you the assistance of visual remembrance.

If you should refer to your dictionary for the spelling of a word, write it once or twice correctly by the side of a misspelled copy, and compare the two modes, pronouncing the word aloud. When next you require to write this word, your tongue, ear, eye and hand will conjointly aid you.

If you are going to commit to memory a long piece, write out a small portion at a time, and carry it about with you, looking at it whenever opportunity offers. Many persons, acting upon this, have adorned their minds in no mean manner, without ever "sitting down" to study. When walking in the streets, or engaged in minor pursuits, we are apt to waste our time in "thoughts revolving." A slip of paper from the pocket, used as proposed, may remedy this.

Do not wait till you find time to accomplish a great deal, but attempt a little immediately. Learn a small portion daily, and occasionally repeat, in suitable divisions, the whole of that which you have learned. The latter injunction should not be neglected, as it is quite as important to retain, in available condition, the results of past application, as it is to make fresh acquisitions.

It is an excellent plan to place the piece of composition you wish to learn before you of a morning, when dressing, and learn as you proceed with your toilet. It is also good to repeat the piece just before going to bed; that which is then brought before the mind, though apparently imperfectly known at night, is often found thoroughly mastered in the morning.

When learning by heart, it is well to retire to some room or locality in which you are not likely to be . interrupted, and there repeat aloud. Poetry may sometimes be learned with speed, by putting a well-known tune to it.

STUDY.

Some people learn best seated, others prefer standing or walking; and these last modes are certainly healthier, as the nervous action caused by the effort of learning can then be greatly modified. Sitting tends to cramp the chest and bend the shoulders.

When you feel in a humor for study, be sure you try to keep so. At such times it is highly important to avoid hearty meals and distracting subjects. Many people can learn best when they are rather hungry, and, when their attention is distracted by a gnawing in the stomach, allay such sensations by a small biscuit and a little water. A meal would incapacitate them for further study. As a rule, we habituate ourselves to eating much more and oftener than we require; and while we are engaged in close mental pursuits, the physical wear and tear is not so great as when we are more vigorously employed, consequently less food is required; a surplus supply exhausts the energies, instead of renewing them.

The time selected for study will greatly influence our success. Some can do best before breakfast, while others appear to wake up intellectually at night, and can continue to study till morning. It is always unwise to attempt hard study immediately after a hearty meal. To persist in so doing will eventually muddle the brain, impair the digestion, and injure the general health.

These remarks are particularly worthy of the attention of those engaged in business. With many, the evening is the only time they get to themselves; and it is highly important to know how to use it to best advantage.

Don't force your attention when you are

weary. This is an invaluable advice; but, unfortunately, it is very difficult to avoid the necessity of violating it. In fact, to work when tired is the duty of most students. It then becomes a matter of importance to know how to meet this demand most judiciously. When the mind has been poring over some abstract subject for hours, without making much headway, a restlessness is experienced. First one foot may be sent in one direction, then the other in another; then the arms fall listlessly by the sides, and the performance is completed by a yawn. How you long for a few winks! Well, close your eyes, lose yourself and wake refreshed. Get some one to rouse you when you have dozed a few minutes, for fear you should prolong your sweet oblivion. This is also very effective in case of grief and anxiety.

Another excellent method is to rise, if you are seated, to stretch yourself, close your hands, and strike out vigorously, right and left, for a few moments.

Washing the face and hands in cold water is also good; or simply applying a damp towel to your eyelids.

Sitting near the fire will tend to draw you off to sleep, and breathing hot or impure atmosphere will make you drowsy; so do not make yourself too cozy, and see well to the ventilation.

A short run in the open air, or rushing up and down stairs, will be found reviving. Those confined indoors during the day should take a few minutes walk in the fresh air of an evening, before commencing to study.

When reading produces headache, it may often be removed by passing the tips of the fingers of both hands a few times from the centre of the forehead to the commencement of the cheeks, either at the distance of about of about half an inch from the face, or in contact with it. Many have tried this with success.

Some students suffer greatly from thirst. By gargling the throat, and rinsing the mouth with cold water, this feeling may be removed; some dip a crust of bread in cold water and place it in their mouth. This may be tried by "thirsty souls" as well as students.

When you wish to concentrate your mind upon a given subject, prevent, as far as possible, anything from distracting your attention. If you have several things to do, always execute those of importance first, as a most powerful cause of distraction is the knowledge of duty unperformed.

Many people lament the amount of time lost

in traveling, as they cannot then study or read, except at the penalty of a headache or indisposition.   They should repeat mentally that which they have previously learned.   Some can read very well, when traveling, by placing a card immediately below the line upon which the eye is fixed; this greatly counteracts the effect of the oscillation of the carriage.

Change of thought is most refreshing to the mind.   When close application has wearied you with a subject, turn to a fresh and more interesting one.   Change of work is rest.

Let it be distinctly understood that I do not advocate unnecessary irregularity either in diet, exercise or sleep, but simply give certain hints that may prove advantageous when irregularity cannot be avoided.   Cultivate regular habits as much as possible.   Over study and mismanagement will undermine the best constitution. Use wise moderation in everything; few have an idea of the amount of application which can be endured by a careful observation of nature's laws.

CARELESSNESS.

That which is commonly regarded as defective memory, is in many instances simply the result of carelessness, which may be manifested by—

Want of attention;
    "   " system;
    "   " reflection;
    "   " promptness.

Endeavor to bring your mind solely on the subject you have in hand. Observe thoroughly what, when, where, how, why, etc.

Be systematic; keep everything in its proper place, and you will save your time and patience when you look for the required article. Accustom yourself to give a glance or thought to necessary requirements, before proceeding about your business.

Be prompt; do not defer until to-morrow what should be done to-day.

Carelessness is often productive of nervousness and anxiety. The person who lacks attention, system and promptness, is repeatedly seized with a panic—fearing that something has been forgotton, or mislaid, or omitted—while the careful man, knowing that everything is in order, can look events calmly in the face.

ATTENTION.

Attention is an important preliminary to memory. Unless we pay attention to what is going on, it is impossible to remember anything about it. The two great causes that tend to destroy attention are, External Impressions and Internal Emotions. Thus, suppose a person is nominally listening to a lecture or conversation, he may positively not be hearing it at all; either his attention has been attracted by something he has just seen, and he may be thinking of that; or something said in the address may have aroused a host of associations, may have caused a train of thoughts, and led his mind astray. And mark, if the address is inanimate, dry and without ideas, the distracting influence of external impressions will be felt by many listeners; if the discourse is spirited and interesting, the internal emotions will act just as powerfully with others. Well, what can be done to fix the attention? We must establish a mnemonical counter-attraction—fortify our minds with our attention-taking, thought-securing basis of association, upon which we can arrange and fix our ideas.

There are several other minor causes of distraction: an unconscious liking for reverie; an indiscreet supply of food, and the existence of bad ventilation. The latter may often be remedied by opening a door or window, but the other conditions rest with the individual. Certain articles of diet upset some people, or make them drowsy; and over-eating is almost sure to have the same effect. Some people appear to think a church the legitimate place in which to let the head go wool-gathering. They willingly allow the mind to wander to the ends of the earth, and then express surprise at not remembering the sermon. Reverie is a species of mental dissipation which is, in the length, prejudicial to the development of intellectual power, and those who wish for improvement, should strive to conquer the habit. The eyes have a great deal to do with attention. If we look about us, we are likely to see many things which may distract us; so it is desirable to keep the eyes on the speaker, if possible, not on bald heads or pretty bonnets—now and then a voice should whisper: "Pay attention!"

MENTAL PICTURING.

Perfection of association is that which secures the harmonious action of the greatest number of powers which can be brought into use for the object desired. We may fail from want of articulation, but more frequently forgetfulness arises from imperfectly picturing. Impressions may be made variously, sometimes thus: The tongue gives an utterance which is conveyed to the ear, the ear-received utterance produces a mental picture which is received by the mind's eye, and the impression on the eye awakens reflection or mental comment. It often happens that the remark made on a thing is better remembered than the object itself. Hence the importance of reflection as an aid to memory.

A witty remark is often well remembered. A pupil could not tell which arm Nelson, the great English Admiral, had lost, but, on being informed, said, "I shall not forget that now; I see it was not the one which was *left*."

In the ordinary way it sometimes happens that things are forgotten, because they can only be seen mentally, or thought about with great difficulty. Artificial memory here supplies the means of rendering them retainable. In remembering, it would be well not to tax one or two particular powers, but to secure an agreeable division of labor by means of artificial memory aids.

ARTIFICIAL MEMORY.

There are many natural operations of the mind, which, when properly understood and performed, assist the memory wonderfully. Some of these have long been in use under the names of Mnemonics (derived from a Greek word, meaning to remember), and Artificial Memory, but, owing to the short-sighted policy of the teachers of these methods, who exact a promise of secrecy from their pupils, the systems are not much known to the general public. Books on Memory have certainly been published, but the greater part of them are mere advertisements of methods taught by their authors, and contain nothing new.

Men of genius and high standing have praised and advocated artificial memory aids, and a host of living scholars recommend their use, yet there are persons who, without even investigating the merits of the system, or from thoughtlessness, declaim against the adoption of artificial aids. This is rather amusing, when we consider that reading, writing and printing are but artificial

means of presenting ideas to the memory.
What is the written word "memory?" It is an
artificial combination of artificial signs, which,
by common consent, represents to our memory
that particular power of the mind artificially
named "memory."

The antagonists of mnemonics reason to this
effect. To bring ideas before the memory by
the artificial means of reading, writing and
printing, is wise; but to bring the same ideas '
before the memory by any other artificial mode,
is folly. Such able reasoning is, perhaps, too
profound to be understood in our unenlightened
times, but as we advance, we shall no doubt see
more clearly.

Others, again, will not listen to the facts that
proclaim artificial memory one of the easiest
arts to learn, and to apply, but state that they
cannot spare the time to master it, or that they
will not burden their heads with any more lore.
The man who is in a hurry to proceed to a dis-
tant place, and is offered a fast horse to take
him there, might just as well say, "I have no
time to mount the horse," and walk on. Or the
man who has to dig a hole in a rocky ground,
and is offered a pick to assist him, might with
equal reason, reply "I can't burden myself
with a pick," and use his fingers. Of this latter

class was the man who, on going to visit his friend, and not caring to think of the number of the house where his friend lived, because it would bother his head, went to every house in the street to enquire for him.

The anti-mnemonist and the other man, will find the following extract from the "Persevering Student" applicable to their case.

"Seeing that it was highly important to make haste, he delayed as much as possible; and having great faith in early rising, he lay in bed till twelve, and sat up late. Knowing that he possessed a remarkably sieve-like memory, he poured knowledge into it as fast as possible, in order to obtain a full mind. Desiring to study History, he at once commenced with Hebrew. At the examination, the results were most satisfactory, for, on finding himself plucked, he found consolation in the thought that he had not been idle, and decided upon devoting another year or two to similar study. He soon fell into indolent habits, however, and became seriously ill from over application; but, as he may recover before he dies, we will leave him here."

Many people use artificial means of aiding the memory, peculiar to themselves. The country lass will often tie a knot on her handkerchief, or a string on her finger, to be reminded of some errand or engagement. The city dame will often aid her memory in remembering the number of a house, by selecting words containing the requisite number of letters to represent the figures. No 66 would be recalled by "finest houses"—both words having six letters; 34 = "bad girl"—*bad* having three, and *girl* four letters. These are, however, poor methods.

The mnemonist uses surer and more systematical means. Give any person unacquainted with artificial memory a hundred disconnected words or figures to learn. He will find it a long and difficult task, and, unless his memory is exceptionally good, he will, after all, feel uncertain whether he can remember them all or not. Give the same task to a mnemonist, and he will grapple with it like a strong man rejoicing in his power. He will find the work easy, and certain of being accomplished.

The advantages resulting from the use of this system, aside from its means of aiding the memory, are numerous. It improves the intellect generally, enlarges the imagination, gives confidence, and diminishes the probability of overtaxing the faculties. It strengthens the judgment by enabling a person to keep more facts and knowledge before the mind. It induces a systematic mode of thinking and acting: the

man who uses no method in remembering, holds
facts in his head in a state of confusion, and
does not know where to find them; while the
mnemonist puts each in its place, and can produce them whenever they are wanted. It saves
time: the student who learns in one hour, by
artificial memory, what under ordinary circumstances would have taken him five hours or more
to master, has saved the difference in time.
This spare time he may devote to further studies, or to exercise. The classmate who strains
his brain by useless repetitions, and languishes
for want of the necessary exercise to refresh his
head and body, sees with envy and astonish-
ment that his comrade, who is considered to
have a poorer head than himself, not only sur-
passes him in accuracy and amount of knowledge, but finds time for recreation. It serves
to fix the attention: a sprightly, interesting
subject is sure to secure the attention; but,
unfortunately, there are a great many abstract
and dry matters that will, in spite of our best
efforts, send us "wool-gathering." The pleasant
mode of using artificial memory will greatly
counteract this by making the subject more
lively. Further, by enabling us to dispose of
the ideas and facts presented, we find ourselves
in a position to pay more attention to minor, or
to succeeding details.

ARTIFICIAL MEMORY.

CHAPTER I.

ASSOCIATION.

Association of ideas is the principal agency of artificial memory, and its motto: Link thought with thought, and think of the impression. The power of association to call up forgotten circumstances must be well known to every one. The knot on the handkerchief, or the string on the finger, serves to recall the errand or engagement that might otherwise have escaped the memory. What a host of remembrances are brought to life by a visit to the home of your childhood! Every house, every view, appears loaded with incidents that perhaps never entered your thoughts from the time you left their scene until you now return, years older and wiser. You approach the moss-covered old oak, and notice in the bark a nearly overgrown mark that recalls some act of boyish mischief, or, to be more charitable, some memorial. You wander by the stream, and a tiny "forget-me-not" meets your eye. Instantly there rush through your memory the scenes of the day when, with a voice trembling from emotion, you offered the flower to one who was dearer to you than everything else. A song, a glimpse, a mere trifle, will bring old ideas fresh before you. In company, where the conversation is pretty diversified, you may often hear persons illustrate this association by exclaiming, "Ah, that reminds me!" some remark having roused a slumbering incident. This serves to show that the well known tends to call up the less known.

Artificial memory is the art of using this association in a systematic manner: to connect with well known facts those that we wish to remember, so that by referring to the known facts we may be able to recall the less known ones connected with them. To form mental pictures is the only exertion required of the student, and will be found very pleasant.

When there is no connection between the words to be committed to memory, it is difficult to retain them; we must, therefore, establish a connection, and link one word with another. Try to remember the following words by reading them over once only:

| cow | Mississippi | bridge |
| storm | coat | apples |

pins	legs	tree
house	cellar	woman
	stars	life

You will probably find it difficult. Well, form a connection; link the words together, taking care that you form a clear, vivid picture —one that you will thoroughly realize in your mind, before you leave it. By a little practice you will acquire a wonderful facility in picturing. You may link the words as follows:

A *cow* swimming across the *Mississippi* (imagine that you actually see a cow swimming and splashing in the river) because the *bridge* had been destroyed by a *storm* (fancy that you see before you a bridge ruined by a storm, and consider this the reason for the cow's swimming the river, instead of walking across), wore a *coat* (fancy a cow with a coat on), which was full of *apples* with *pins* stuck in them (imagine that you see this); in running, the apples knocked against her *legs*, which so frightened her that she ran up against a *tree* (picture) growing by the side of a *house*, and tumbled down into the *cellar*, falling right upon a *woman*, and causing her to see more *stars* than she ever saw in her *life* before (imagine that you are standing near by, and observing these curious incidents).

Connecting the words in some such absurd manner, you will find no difficulty in naming them as you mentally review the picture formed.

Try another list of words:

man	fop	money
pen	summer	coal
hat	tanner	eggs
window	crowd	stones
palace		river

And connect them somewhat like this: A *man* who looked like a *fop*, and made *money* by his *pen* in the *summer*, dropped a red-hot *coal* upon the *hat* of a *tanner* who was eating *eggs*, while looking thro' the *window* at the *crowd* that was throwing *stones* over the *palace* on the other side of the *river*.

Reflect upon each picture before you proceed to the next one. Unless you make the picture vivid, you will not remember the words so well; you must imagine that you are an actual spectator of the scenes formed in your picture, or an actor in them. Write down a number of words at random, and form your own connection.

A more effectual plan of committing words to memory is to connect them with a well-known piece of poetry or prose. One containing plenty of nouns is best. Say that your list of words commences with, *leaf, fire, giant, crystal, horse,* etc., and that your poem begins thus:

> "Under a spreading chestnut-tree,
> The village smithy stands;
> The smith, a mighty man is he,
> With large and sinewy hands."

Take the first noun in the poem, *chestnut-tree*, and connect *leaf* with it. "A *chestnut-tree* with but a single *leaf*," will do for a picture. *Fire* is the next word, to be connected with *smithy*. Imagine a roaring *fire* in a *smithy*. "A *smith* as tall as a *giant*," will do for the following picture. Connect *man* and *crystal*: A man swallowing a crystal. *Hands* and *horse:* A horse born with hands instead of hoofs, and so on. Having formed a vivid picture of each word in connection with the nouns in your poem, you can recall them by thinking of the words in the verses. *Chestnut-tree* will recall *leaf; smithy* will suggest *fire; smith, giant,* etc.

Simple and paltry as this may at first appear, you will find it of great value.

Following chapters will contain more complete and methodical plans of remembering words and ideas.

CHAPTER II.

MNEMONIC ALPHABET; MEMORY TABLE.

Before we proceed any further with the system for remembering words and ideas, we must learn a plan by which to recollect figures. It is far easier to remember words than figures. You will not be so likely to forget the word "street," as the number 6,232, or the name "Franklin," as the figures 539,079; and yet there is a way by which figures can be remembered as easily as words: by translating figures into letters, numbers can be formed into words.

The first step is to learn the Mnemonic alphabet. A careful study of different systems has convinced the author that the alphabet here given is superior to any which he has seen.

Let

j, and the similar sounds *g* (soft), *sh*, *ch*,	represent *1*	Capital J is like a 1; it has one limb.
t, and its subdued sound *d*,	" *2*	The letter *t* is the principal sound, and only consonant, in the word *two*.
r,	" *3*	The letter *r* is the principal sound in the word *three*.
m,	" *4*	Capital M has four strokes.

f, and its subdued sound *v*,	"	**5**	The letter *f* and *v* are both found in the word *five;* V is the **Roman** figure for 5.
s, and the similarly sounding letters *c* (soft) *z*,	"	**6**	The letter *s* is the chief sound in the word *six;* capital written *C* looks like a 6.
l.	"	**7**	Capital *L* inverted, is like a 7.
b, and its sharp sound *p*,	"	**8**	Capital B is like an 8.
n	"	**9**	The letter *n* is the chief sound, and only consonant, in the word *nine.*
q, and its similar sounds *k*, *c* (hard), *g* (hard), and *ng*.	"	**0**	Capital *Q* is like a 0; *c* is also like it.

Vowels are not reckoned; neither is *w* nor *h*, on account of their being silent in many words.

It must be borne in mind that we always go by sound, and not by spelling. The word *calm* we pronounce as if spelled *kahm; cough,* as if spelled *kof; night* as *nite; giddiness* as *gid-i-ness; notion* as *noshun,* etc. The reason for this is that, when a figure has been translated into a word, and you wish to reconvert it, you may have no trouble about the spelling of the word. Some might forget how *knight* is spelled, and get confused when trying to translate it into a figure. Phonetic spelling prevents the possibility of mistakes.

Let us fix the alphabet firmly in memory by going over it once more. The figure 1 can be translated into *j*, as in *joy;* into *g*, as in *George;* into the subdued sound *sh*, as in *shade, sure, notion;* into the sharp sound *ch*, *tch*, as in *chair, ditch.* The figure 2 is represented by *t*, its kindred sounds *d* and *th*, as in *that, third.* The figure 3 is represented by *r*, as in *rower.* The figure 4 is represented by *m*, as in *mama.* The figure 5 is represented by *f*, and its heavy sound *v*, as in *favor.* The figure 6 is represented by *s*, *c* (soft), and its hissing sound *z*, as in *cease, zero, wiser.* The figure 7 is represented by *l*, as in *lily.* The figure 8 is represented by *p*, and its heavy sound *b*, as in *public.* The figure 9 is represented by *n*, as in *noon*, and, lastly, the cipher (0) is represented by *q*, *k*, *c* (hard), the heavy sound *g* (hard), and the nasal sound *ng*, as in *coke, quill, gong.*

The next thing is to translate words into figures, and figures into words, for practice.

Arkansas = 30966. The *a*, being a vowel, is not reckoned; *r* stands for 3; *k* = 0; *a* is not counted; *n* stands for 9; *s* = 6; *a* is skipped; *s* = 6; *rk-ns-s* = 30966. *Mississippi* = 4668; *m* = 4; *ss*, equal to one *s* sound, stands for 6, $ss = 6$; $pp = p = 8$; m-ss-ss-$pp = 4668$. *Knowledge* $= 971$; k is not sounded; $n = 9$; *ow* is not counted; $l = 7$; *dg* sounds as *j*, which equals 1; n-l-$dg = 971$. *Defile* $= 257$; *remove* $= 345$; *homely* $= 47$; *housewife* $= 65$; *Californian* $= 075399$, etc.

Continue to practice in this way for a short time, either on paragraphs in this book, or in the head, and then try to form words out of figures. For instance, 172 may be rendered into *child, shield* or *jollity;* $547 = family, half-a-mile, foamy ale;$ $369 = raisin, our son, war sign;$ $830 = prong, pork, brig;$ $83626 = parasites, priestess, poor seats;$ $57496 = full moons, a few lemons, vile means,$ etc.

Facility in the translation of figures into words, and *vice versa*, is acquired by just a little practice. If you have to translate, say 57, you take the representing letters *f* or *v* and *l*, and, commencing with the vowels in their order, a, e, i, o, u, y, run over them quickly in conjunction with the consonants, like this: $f a l = fall;$ $f e l = fell, = feel;$ $f i l = fill;$ $f o l = foal, = fool;$ $f u l = full;$ and then $r a l = rale;$ $v i l$ = no word, $v y l = vile;$ again, $h o v l = hovel,$ etc. Running them over like this, you can stop at the most appropriate word, and fix upon that to represent the figures. When you have formed the figures into words, you should associate them with each other, or with some piece of poetry, etc. by making a vivid picture; you will then be able to remember figures with as much ease as words.

There are many cases, however, when it will be found difficult to form a good picture of any length, or, when several pictures have been made, to recall the leading words that will suggest the other parts; in many instances you may have one word only to remember, and require some permanently fixed picture words to connect with. The following table of words suitable for picture-making will supply this want. These words are to be used as "memory-pegs," on which to hang facts, ideas or words, or as "pigeon-holes," into which to place them; so that, by referring to any one of them, you may find the facts attached.

The table is composed of words representing the numbers from 1 to 100, and may be extended by the pupil as occasion requires. If the figure alphabet has been mastered, you will find no difficulty in committing the table to memory by reading it over carefully once; and once learned, it will serve through a lifetime as a means of remembering, with ease and certainty, a thousand matters which may be desirable to know.

MEMORY

0 Key	1 Shoe	2 Toe	3 Hair	4 Ham
10 Jig	11 Judge	12 Shot	13 Chair	14 Jam
20 Dog	21 Ditch	22 Tooth	23 Door	24 Dome
30 Rock	31 Wretch	32 Rod	33 Warrior	34 Room
40 Mug	41 Match	42 Mouth	43 Moor	44 Mummy
50 Fig	51 Fish	52 Fight	53 Fire	54 Foam
60 Sack	61 Sage	62 City	63 Czar	64 Swim
70 Lock	71 Leech	72 Lad	73 Lyre	74 Lamb
80 Pick	81 Bush	82 Boat	83 Bear	84 Beam
90 Neck	91 Niche	92 Naiad	93 Norway	94 Name

TABLE.

5 View	6 Saw	7 Hill	8 Bee	9 Hen
15 Shave	16 Cheese	17 Jail	18 Ship	19 Chain
25 Dove	26 Dice	27 Towel	28 Tub	29 Den
35 Roof	36 Rose	37 Rail	38 Rope	39 Rain
45 Muff	46 Mouse	47 Mule	48 Map	49 Moon
55 Fife	56 Face	57 File	58 Fop	59 Fan
65 Safe	66 Sauce	67 Seal	68 Spy	69 Sun
75 Leaf	76 Lass	77 Lily	78 Lip	79 Lane
85 Beef	86 Bus	87 Pill	88 Puppy	89 Pin
95 Knife	96 Nose	97 Nail	98 Nap	99 Nun

With the table all the drudgery of the system is surmounted; you have only to apply what has already been learned. Feats of memory, which the world at large would consider beyond human power, will now be easy to perform.

Key will do for 0; *k*, the only consonant in the word, representing 0, according to the alphabet. *Shoe* will do for 1; the only consonent sound, *sh*, representing 1. *Toe* stands for 2 ($t=2$); *Hair*$=3$ ($r=3$); *Ham*$=4$ ($m=4$); *View*$=5$ ($v=5$); *ditch* $=21$ (d-$tch=2$-1); *safe*$=65$ (s-$f=6$-5); *nail*$=97$ (n-$l=9$-7). And so on. The words in the table being formed by a translation of the figures against which they stand, will at once be understood and remembered, and may be extended beyond 100 by a similar process of translating.

Since it is necessary to have the table, or the first part of it, run fluently on the tongue, in order that there may be no hesitation when a "peg" is wanted, it would be well, if only for practice, to form a series of pictures of the first twenty or thirty words, which will be most in use. You might commence thus:

Finding a *key* in my *shoe*, under the big *toe*, I enveloped it in *hair*, and hid it in a *ham*, which I placed on *view* by the old *saw* on the *hill*, where a *bee* and a *hen* were dancing a jig, etc. You can picture the words thus: 0 stands for *key*—the handle of a *key* is like a 0; 1 stands for *shoe*—a one-legged man has only *one shoe;* 2 stands for *toe*—we have two big *toes*, or *two* sounds like *toe;* 3 stands for *hair*—*hair* is like a bushy *tree*; 4 stands for *ham*—imagine a big *four* written on it; and so on.

This mode, however, of learning the table, not one in twenty will require to use; for the translation of the figures, aided by a vivid picturing of the word, will serve to fix it in memory.

CHAPTER III.

HOW TO REMEMBER WORDS.

Having mastered the table, or as much of it as you chose to learn, you will be shown how to apply it, so as to remember any number of words after reading or hearing them once. Take these words as an example:

guy	plank	boat
magpie	battle	mud
sea	thistle	rum
pole	sheep	tea
reef	woman	pig
candy	bottle	silk
crystal	carpet	gravel
iron	window	fire
stocking	estate	street
rope	tiger	frog
palace	boy	horse
flowers	cloth	ear
stream	cottage	bread
monkey	powder	mountain
knife	promenade	hat
	soap, etc.	

Connect each word with one of the pegs in the table, by means of a picture. The first word in the list is *guy*; connect it with peg 1,

which is *shoe*. In this illustration we skip peg
0, for reasons given below. Make a vivid picture of it; use your imagination. Do not be
particular with regard to the character of your
picture; the more absurd it is, the better you
will remember it. A ridiculous thing will strike
the mind more than any ordinary event. The
pictures here given as an illustration are the
first that suggested themselves to my mind, and
the pupil can, no doubt, form better ones. But
mind, do not *read* the words of your picture
merely; imagine that you actually *see* the scenes
described, that you take a part in them.

Associate *guy* with *shoe*. Fancy a *guy* with only
one *shoe*. Next *plank* with *toe*. Imagine
that you get into a rage, and kick your *toe*
through a thick *plank*. Next follows *boat*,
to be joined to *hair*. Picture to yourself a
boat, loaded with human *hair*, sailing past.
Magpie and *ham:* a thievish *magpie* flying
away with your breakfast *ham*. *Battle* and
view: a *battle* of which you have a *view* from
an eminence. *Mud* and *saw:* a *saw* sticking
in the *mud*. *Sea* and *hill:* a green *hill* in
the middle of the *sea*. *Thistle* and *bee:* a
bee trying in vain to get sweet from a *thistle*.
Rum and *hen:* a *hen* getting drunk on *rum*.
Pole and *jig:* a man dancing a *jig* on a *pole*.

Sheep and *judge:* a *judge* sitting on a *sheep.*
Tea and *shot:* on opening a packet of *tea* to
find it adulterated with *shot.* *Reef* and *chair:*
a *chair* washed on to a *reef.* *Woman* and
jam: a *woman* making *jam.* *Pig* and *shave:*
a *pig* running into a barber shop to get a
shave. *Candy* and *cheese:* a *cheese* made of
candy. *Bottle* and *jail:* the *bottle* is the chief
supporter of the *jail.* *Silk* and *ship:* a *ship*
with sails of *silk.* *Crystal* and *chain:* a *chain*
of *crystal.* *Carpet* and *dog:* a *dog* tearing up
a *carpet.* *Gravel* and *ditch:* a *ditch* filled up
with *gravel.* *Iron* and *tooth:* insert a *tooth*
of *iron.* *Window* and *door:* a house without
window or *door.* *Fire* and *dome:* the *dome*
on *fire.* *Stocking* and *dove:* a dead *dove* in a
stocking. *Estate* and *dice:* losing an *estate*
on *dice.* *Street* and *towel:* flinging the *towel*
into the *street.* *Rope* and *tub:* a *tub* hanging
by a *rope.* *Tiger* and *den:* going into a *den*
and falling on a *tiger.* *Frog* and *rock:* a live
frog in a *rock.* *Palace* and *wretch:* a *wretch*
breaking into the *palace.* *Boy* and *rod:* a
boy balancing a *rod* on his chin. *Horse* and
warrior: a *warrior* getting his *horse* killed
under him. *Flowers* and *room:* a *room*
dressed with *flowers.* *Cloth* and *roof:* a *roof*
made of broad-*cloth.* *Ear* and *rose:* wearing

a *rose* in her *ear*, instead of an earring. *Stream* and *rail:* crossing the *stream* on a single *rail*. *Cottage* and *rope:* an old *cottage* held togetter by a *rope*. Two or more words may be connected with one peg, without extra trouble. Associate *bread*, *monkey* and the peg *rain*. A *monkey* moistening his *bread* in the *rain*. *Powder, mountain* with *mug:* a *mug* of *powder* blowing up à *mountain*. *Knife, promenade, hat* with *match:* a man taking a *promenade*, with a burning *match* and a bare *knife* stuck in his *hat*. *Soap* and *mouth:* slipping a piece of *soap* into your *mouth*, instead of sweetmeat.

Your list being committed to memory by such picturing as this, you have only to think of the pegs, which you can always do, and they will at once suggest the word or words associated with each of them.

shoe	recalls	guy
toe •	"	plank
hair	"	boat
ham	"	magpie
view	"	battle
saw	"	mud
hill	"	sea
bee	"	thistle
hen	"	rum

jig	"	pole
judge	"	sheep
shot	"	tea
chair	"	reef
jam	"	woman ·
shave	"	pig
cheese	"	candy
jail	"	bottle
ship	"	silk
chain	"	crystal
dog	"	carpet
ditch	"	gravel
tooth	"	iron
door	"	window
dome	"	fire
dove	"	stocking
dice	"	estate
towel	"	street
tub	"	rope
den	"	tiger
rock	"	· frog
wretch	"	palace
rod	"	boy
warrior	"	horse
room	"	flowers
roof	"	cloth
rose	"	ear
rail	"	stream

rope	"	cottage
rain	"	bread / monkey
mug	"	powder / mountain
match	"	knife / hat / promenade
mouth	"	soap

This shows that by means of the table you can instantly remember any number of words, ideas or facts, in any order that you wish. You may repeat them backwards or forwards, tell the 25th word, the 15th, the 87th, or any one required. If you are asked to name the 28th word, you think of (2-8=t-b=) *tub*, which recalls *rope*. The 42d peg is *mouth* (4-2=m-th), which recalls *soap*. If you wish to know what number in the list *fire* is, *fire* will at once recall *dome*, which is the 24th peg. Fire is therefore the 24th word in the list. Had we included peg 0 in the picturing, we would have had to add 1, in order to arrive at the number of the words; by commencing with peg 1, this is avoided.

CHAPTER IV.

HOW TO REMEMBER FIGURES.

A preceeding chapter has already explained the method by which figures are translated into words, so as to be more readily retained. Suppose that you wish to remember the number of a house, say, 238 Market street; by translating 238 into the letters *t* or *d*, *r*, *b* or *p*, and filling in the vowels, you can form the words *tribe*, *trap*, *trip*, *troupe*, *drop*, *dear boy*, *tow-rope*, etc., and associate one of them with Market street: taking a *trip* through the *market*; a *troupe* performing in the *market*, etc. Or, in case the whole picture should slip out of memory, connect it with a peg, thus: *key* lost in a *trap*, while passing through the *market*. *Key*, which you can always think of, will recall the words associated with it, viz: *Market*, which you know means Market street; and *trap*, which you will translate into the number.

You may translate 293 into *tanner;* 176 into *jails;* 458 into *move up;* 3681 = *rosebush;* 8760 = *pull a scow;* 5491 = *half my niche;* 13940 = *a cherry in a mug;* 572359 = *flighty raven;* 192 = *chant;* 796 = *lines;* 248 = *damp;* 663 = *saucer;* 349 = *ermine;* 696 = *sins;* 350 = *roofing;* 486 = *maps.*

Let us take a list of fifty-one figures, and see if we cannot learn them in a very short time:

363 273 806 617 323 642 591 700 384 906 157 386
756 613 942 478 320.

They seem pretty formidable as they stand, but by applying the system, the task becomes easy, and would be so if the list were twice as long.

Separate the figures into groups of three, by pencil strokes. Then translate each group into a word or words, and apply your memory pegs to each group in its order. By arranging the figures in groups of three each, you will be able to name the 25th, the 12th, the 50th, or any one required; if this is not desired, you may form words representing one, two or more figures, as they happen to suggest themselves:

The first group, 363, may be rendered into *horse-hair*, and connected with peg 1; a *shoe* stuffed with *horse-hair*. The second group, 273, may be formed into *tiller*, and joined to *toe;* the captain holding the *tiller* with his *toe*. Make the picture vivid; imagine that you actually see the things described!

Next, 806 = *books*, to connect with *hair*; some *books* tied together by a single *hair*. 617 = *satchel;* a boy carrying a *ham* in his *satchel*. 323 = *writer;* a *writer* describing a romantic *view*. 642 = *smith;* a blacksmith using a *saw* on a piece of iron. 591 = *funny show;* a *funny show* taking place on a *hill*. 700 = *oil-cake;* a *bee* sucking an *oil-cake*. 383 = *robber;* a *robber* stealing a *hen*., 906 = *nags;* some miserable looking *nags* dancing a *jig*. 157 = *shovel;* a *judge* reduced to work with a *shovel*. 386 = *robes;* the kingly *robes* riddled with *shot*. 756 = *leaves;* grandfather's *chair* adorned with green *leaves*. 613 = *sea shore;* a boy eating *jam* found on the *sea shore*. 942 = *new mode;* I discover a *new mode* to *shave*. 478 = *my lip;* holding a big *cheese* on *my lip*. 320 = *red cow;* a *red cow* tearing down the *jail*.

You have now only to think of the peg, which will be sure to recall the figure in the picture. *Toe* will recall *tiller*, which stands for 273 (t = 2; l = 7; r = 3). *Judge* will suggest *shovel* = 157. If you are asked to name the eleventh figure, which is the middle figure of the fourth group, you think of peg 4, *ham;* this will suggest *satchel* = 617, of which 1 is the middle figure. The forty-fiftb figure (divide 45 by 3), is the last figure in group fifteen, peg 15, *shave* will recall *new mode* = 942, of which 2 is the last figure.

This process may appear a little complicated at first, but, if you know the alphabet well, it ought to be very easy. Practice will make you expert, besides greatly benefiting your imaginative and other faculties.

4

CHAPTER V.

HOW TO REMEMBER DATES.

To recollect dates has hitherto been a dry and difficult task to the student. By following the method here given, he will find it transformed into an easy and interesting one.

In trying to bring to mind some line of poetry, its first word, syllable, or even letter, is often found sufficient to recall it. This fact is largely made use of in remembering dates. Of the first word, syllable or letter of the name of the event to be remembered form a phrase or word that will suggest the event, and at the same time bear some relation to another word or phrase formed by a translation of the date.

If you wish to remember the date of the discovery of America, 1492, and the name of the discover, Columbus, you may do so in this way: America has a *mount* that looks like a *column; America* will recall the picture of the *mount*=492, looking like a *column*, which will suggest *Columbus.* In modern chronology, there is no necessity for noticing the thousand, therefore 492 will be sufficient to recall the whole date, 1492. Again, " America made independent, 1776," could be rendered into American *Indians* dislike *lilies*. The first syllable of *Indians* will suggest *independence*; *lilies* = 776 = 1776.

When you have a series of dates to learn in a given order, you may connect them with the pegs, taking care to make the relation between the peg, the event, and the date as close as possible, in order that the impression may be retained.

Should there be several kings of the same name in your lesson, you must indicate which one is meant, either by giving him an attribute, in translating the number into a word, or by giving the name an end-consonant that will stand for the number. Louis the XIV = *Low chum;* Louis II = *Lout*—here *L* will be sufficient to recall *Louis*, while *t* will suggest the II.

In order to illustrate my meaning fully, I subjoin a list of events from the history of England, with their dates, rendered into this system.

Events.	Dates.	Pegs.	Pictures.
Augustine lands	596	shoe	*Augustus* exacting *fines* from all who wore *shoes.*
Danes lands	787	toe	To *dance* on one *toe* is easier than to *leap a wall.*
Egbert crowned	827	hair	A *hairy poodle* breaking the *crown* of an *egg-pot.*
Alfred made king	871	ham	*Ham* would be put to the *blush* to learn the *king's alphabet.*

Oxford University	886	view	*View* of *babies* riding an *ox*.
Massacre of Danes,	1002	saw	*Danes* at *mass* wear a *cock-ade* like a *saw*.
Doomsday-book	1086	hill	A *book* meeting its *doom* in the *gaps* of the *hill*.
First Crusade	1096	bee	*Bees* taking a *cruise* in *canoes*.
A Becket murdered	1172	hen	A *hen* killing a *child* with a *bucket*.
Third Crusade	1192	jig	*Three Crusaders* dancing a *jig* in a *shanty*.
Elizabeth crowned	1558	judge	A *judge* giving a *little bit* to a *fife-boy*.
James I crowned	1603	shot	James the First bearing a *scar* from a *shot*.
Charles I crowned	1625	chair	A *stiff chair* makes a bad *couch*.
Charles II crowned	1649	jam	A *seaman* smearing *jam* on his *chart*.
James II crowned	1685	shave	*Shave* the *soap off Jims'* head.
Wm. & Mary crown'd	1689	cheese	*Will Mary* eat *cheese* with a *spoon?*
Anne crowned	1702	jail	*An oil-coat* got a person in *jail* once.
George I crowned	1714	ship	*Ships all a-jam* in the *gorge*.
George II crowned	1727	chain	A *little chain* may be *good*.
George III crowned	1760	dog	A *dog* trying to *gore* a *lazy hog*.
George IV crowned	1820	ditch	A *boy took* a *gem* out of a *ditch*.
William IV crowned	1830	tooth	William took a *whim* to *break* his *tooth*.
Victoria crowned	1837	door	*Victory* sticking in a *barrel* at the *door*.

By reading the pictures you will at once discover the meaning of the different words. Take,
for instance, the picture "To *dance* on one *toe*

is easier than to *leap a wall*." Here the peg *toe* will recall *dance*—suggesting Danes—and *leap a wall*, which you will know how to translate into 787, the date of their landing. In picturing "Oxford University founded 886," you take your peg, which here happens to be *view*, and connect with it the most appropriate translation of 886 (babies, puppies, papers, etc.), as well as some name that will suggest Oxford. "*View* of *babies* riding an *ox*," is the picture that occurred to me. A *judge* giving *a little bit* to a *fife-boy*. The peg *judge* recalls *a little bit*—suggesting Elizabeth, as pronounced by children—and *fife-boy*—representing 558, which, you will be sure to know, means 1558. *An oil-coat* got a person in *jail* once. *Jail* will call up *an oil-coat*, which will readily suggest Anne, 702; *an* being quite sufficient to indicate Anne, the queen, especially if your list for the occasion is composed of sovereigns only. A *little chain* may be *good*. The first letter in *good* will recall George, and the last indicates his number; *little* will reveal 727.

When a list of kings, etc., has to be learned, it will be found well to form a composition, in which the commencing letter of the verbs or nouns represents the initial letter of the king's name, thus: William I, William II, Henry I,

Stephen, Henry II, Richard, John, Henry III, could be remembered by "a Weak Woman Had Stolen a Hat from a Rich Jew, whose Head," etc.

The student will find it interesting, as well as beneficial, to exercise his ingenuity in preparing the pictures. A little practice gives readiness.

CHAPTER VI.

HOW TO FIND THE DAY OF THE WEEK ON WHICH ANY
DAY OF THE MONTH FALLS, FOR ANY NUMBER
OF YEARS, PAST OR FUTURE.

Take an almanac for any year, and see on
which day of every month the first Sunday falls.
Of the name of each month form a word, which
from its sound will suggest that month, and let
its last consonant indicate the date of the first
Sunday.

For instance, the first Sunday of January,
18?3, falls on the 5th. The first syllable of the
word *Geneva* will suggest January, while its last
consonant, $v = 5$, indicates the date of the first
Sunday. The *f* in *fight* suggests February, and
the last consonant sound indicates that the Sunday falls on the 2nd. The list of the months
can be formed thus:

The first Sunday,	January	5th	= Geneva.
"	February	2nd	= fight.
"	March	2nd	= mart.
"	April	6th	= apes.
"	May	4th	= mama.
"	June	1st	= judge.
"	July	6th	= Julius.
"	August	3rd	= auger.

The first Sunday, September 7th = sail.
     "   October   5th = octave.
     "   November 2nd = note.
     "   December 7th = dell.

You may commit the names to memory by connecting them, thus:

In *Geneva* town there was a *fight* in the *mart* between some *apes*, in which the *mama* of *Judge Julius* took a part with an *auger* and an old *sail*, making them squeal an *octave note* higher, and return to their *dell*.

When you wish to know on what day of the week any day of the month falls, proceed as follows:

From the given day of the month substract the number indicated by the last consonant of the name for that month; from the rest subtract 7, or a multiple of 7, as 14, 21, 28; the remainder will be the day of the week—Sunday being reckoned *no* day, Monday *first* day, Tuesday *second* day, Wednesday *third* day, Thursday *fourth* day, Friday *fifth* day, Saturday *sixth* day.

You may wish to ascertain on what day of the week the 19th of April falls. The name for April is *apes*, of which the last consonant $s=6$. Take this 6 from the 19th, and 13 remain; subtracting 7 from this, leaves $6 =$ Saturday, which is the *sixth* day, as stated above. The 19th of

April is therefore a Saturday. Take the 29th of October. The name for October is *octave*, the last consonant $v = 5$. Substracting this from 29th, leaves 24; from this deduct 21—the nearest multiple of 7 to 24—and 3 will remain, which equals Wednesday.

When the date of the day required is smaller than that indicated by the final consonant of the name for the month, add 7, before commencing, instead of subtracting it afterwards. Take 2nd May. The name for May is *mama*, the last *m* indicating 4; as this cannot be substracted from 2, you must first add 7, which makes 9; now deduct the 4, and 5 is left, which shows the 2nd of May to be a Friday.

For dates in 1872, up to February 29th, 1 must be subtracted, before commencing your calculation as above. Previous to that day, subtract 2. For dates in 1871, subtract 3. For 1870, subtract 4. For 1869, subtract 5. For 1874, add 1 before commencing. For 1875, add 2. For 1876, add 3, up to twenty-ninth February; after that date, add 4. For 1877, add 5. These calculations depend upon the fact that the year 1873 is taken as a basis; if any other year is taken, a different addition and subtraction for other years must be made, as the student will find on examining the almanacs.

A shorter mode of remembering the dates of each Sunday throughout the year, is to take the date of the first Sunday of each month, translate into letters, and form a word for every three months. Take the dates above given: January 5th, February 2d, March 2d=522, of which you can form *fated*. April 6th, May 4th, June 1st= 641=*smash*. July 6th, August 3d, September 7th=637=*sorrel*. October 5th, November 2d, December 7th=527=*foothill*. Fix these words by means of a picture, and you will have days and dates at your immediate command.

HOW TO REMEMBER THE NUMBER OF DAYS IN THE DIFFERENT MONTHS.

You may perhaps have heard the piece of rhyme by which schoolboys keep this in mind.

> Thirty days have November,
> April, June and September;
> February twenty-eight alone,
> The rest all thirty-one.

Another mode they have is to count the knuckles. You commence on the knuckle of the first or fore-finger, and say, January. Then you descend into the hollow between the fore and middle-finger, and say, February; up on the knuckle of the middle-finger, March; down into the next hollow, April; up on the knuckle of the third or ring-finger, May; down into the hollow June; up on the last knuckle, that of the fourth or little finger, July. Now return the way you came, commencing on the last knuckle, and say, August; down into the hollow, September; up on the knuckle of the ring-finger, October; down into the hollow, November; and up on the knuckle of the middle-finger, December. You will observe that the months of 31 days all come on the knuckles, while months of only 30 days, and February, fall into the hollows.

The above modes of remembering the number of days in the month are so good, that you will scarcely care for any mental picturing to aid you.

BIBLE TEXTS.

Indicate the chapter and verse by means of letters, and connect them with the name of the book, or part of it.

Moses ii, 7=motley.

Acts v, 11=active judge, etc.

LATITUDES AND LONGITUDES.

Proceed in the same manner as with texts.

STATISTICS, ASTRONOMICAL MAGNITUDES, LOGARITHMS, ETC.

Say that the population of the United States is 38,567,450. Translate this into *rub off a slim fog* or *a rope vessel may have a hog*, and **attach** it to the subject before you, or to a peg. The expenditure for the year is, say, $70,537,680=*a law-giver lies big* or *look for a lazy pig*. Logarithm 34=1.53148, which can be rendered into *your whim=a show for a shampoo*, or *room=watch for a jump.*

CHEMISTRY.

Translate the combining properties, **or the** measurements, into words, and join them to the respective names, or to pegs.

CHAPTER VII.

HOW TO LEARN GEOGRAPHY.

When you wish to impress upon your memory the outline of a country, with its rivers and towns, and their position, etc., trace on the map, or on a transparent piece of paper, a face, or else some animal figure that would seem to suit the appearance of the country. Then proceed to indicate the geographical position in some such manner as the following. A human face is here supposed to be drawn upon a map of the world.

On the centre of the forehead, *you're up* (Europe.)

Before the right ear is *Jam*-aica.

Which *makes sick* (Mexico) an upper part of the right ear.

The right eye is in the Atlantic Ocean.

Between the eyebrows we have the Great Desert, Timbuctoo and the river Niger.

In the left eye we have *new beer* (Nubia),

Which makes a *red sea* (Red Sea).

The front part of the left ear has been struck by a *mad rascal* (Madras), leaving a *seal on* (Ceylon), just below it.

Above the left ear is *the bit of china* (Thibet, China).

South America is on the right cheek bone.

A *guinea* (Guinea) is on the bridge of the nose.

St. Helena is nearly in the *center* of the face. And so on to any extent.

This plan will not only serve to show the form of the country, but securely fix the names of towns, etc., with their position. A boy will learn more geography in an hour by this pleasant method, than he would learn in months by the ordinary hard study.

Another plan is to form a striking composition on the country, in which the prominent places are referred to by puns, or similar sounds, like this:

ITALY.

Italy is *my land* (Mailand, Milan), cries the *tourist* (Turin) who, endowed with *nice* (Nice) *genius* (Genoa), adopts the *lucky* (Lucca) *mode* (Modena) of travelers, passes through the land of the *longbeards* (Lombardy), and enters the *very nice* (Venice) City of the Sea. If enthusiastic, he proclaims Rome as the *type* (Tiber) and *flower* (Florence) of the land; if prosaic, he prefers a *mess in* (Messina) a *sardine* (Sardinia) shop, and a *nap* (Naples) in the shade of the *silly* (Sicily) looking *pillars* (Palermo) that are gaped at by *mariners* (Marino) afflicted with *liver* (Livorno) complaint, etc.

If you wish to retain the names of States in a certain order, take the first or most striking syllable in it, form a word, and connect it with the pegs, or with some piece of poetry. A *man* (Maine) stuffed some *new* (New York) *moss* (Massachusetts) into his *shoe* (peg). Here three States are connected with the peg *shoe*.

GRAMMAR, BOTANY, NAMES, ETC.

Here, again, punning, or the use of words of similar sound, is of great service. If you wish to recollect the meaning of, say, *equinox*, find some word of similar sound to it, and identify the two: *equi*=*equal*, *nox*=*night*. We here suppose that the pupil is ignorant of the language from which this and other words may be derived. Again, *seraph* (the highest order of angels) might be likened to *giraffe* (the highest animal in stature). *Phrenology* finds a similarity in *free knowledge*. *Nemos*=*new moss*. A person's name may be well remembered by linking an appropriate word, suggesting his name, etc., to some observed peculiarity in his dress, manner, occupation or appearance; Mills the banker=a *banker* who got rich by running flour *mills,* etc.   Greeley the journalist=the *growling writer.*

ERRANDS, ENGAGEMENTS, ETC.

Connect the most suggestive word or name in your list with the pegs, or make **one** picture of them.

Get newspaper; order coals; call on Smith; post the letter; buy a bouquet; send telegram; buy boots; get watch mended, etc. One picture could be formed thus:

Taking a *newspaper* filled with *coals* to the blacksmith, I found him reading a *letter* that had arrived in a *bouquet;* this letter contained a *telegram* telling him to buy *boots* for all *watch* makers in town.

The connection with the memory-pegs might be made like this:

Found a *shoe* enveloped in a *newspaper.* My *toe* burned by a red-hot *coal.* The *hair* of the *blacksmith* singed. Receiving a *ham* by the *post. View* of a stage loaded with *bouquets* set in *telegram*-paper. Found a *boot* filled with *watches* in the old *saw*-mill.

POETRY.

You must take a clear view of the writer's description, not passing it over carelessly or attending merely to the sound, but noting the ideas, and picturing as you proceed. You must place yourself as an actor or spectator of the events described. Note the chief or commencing word of each line, couplet or verse, and connect it with your pegs. Take this passage as an illustration:

> When the humid storm-clouds gather
> Over all the starry spheres,
> And the melancholy darkness
> Gently weeps in rainy tears,
> 'Tis a joy to press the pillow
> Of a cottage-chamber bed,
> And to listen to the patter
> Of the soft rain over head.

To insure a recollection of this, you have to imagine yourself an actor in the scene: to fancy that you actually see the *storm-clouds* gathering over the *starry sky* you have been admiring; that it, as a consequence, becomes *melancholy dark*, and *rainy tears* fall *gently* from the clouds; how *pleasant* it now is to *press* the *pillow* in your rustic *cottage-chamber* bed, and how you lie there *listening* to the pitty-*patter* of that *soft rain over* your *head*. As you proceed, you may connect *storm-cloud* with *shoe:* a *cloud* formed like a *shoe.* *Tears* with *toe:* letting hot *tears* fall on your bruised *toe*. *Pillow* stuffed with *hair*, etc.

PROSE.

Observe the important and striking words in each sentence, and connect them with the pegs, while picturing the events as vividly as possible.

LECTURE, SERMON, CONVERSATION, ETC.

Proceed as above, or make short notes on a piece of paper of the chief ideas and points, while connecting the details with that chief word by means of picturing.

HOW TO LECTURE WITHOUT NOTES.

Select the most suggestive words in paragraph or sentence of your projected speech, and connect them with pegs or poetry. You have then merely to follow up the pegs in their order to obtain the matter for your remarks. Clergymen may associate a whole sermon with the words of a chapter before them, or with "Our Father."

When you wish to commit anything to memory permanently, it will be found best to connect it with a piece of poetry or prose with which you are well acquainted, leaving the memory-pegs free for subjects that you wish to remember merely for a short time.

It would be beyond the limits of this small book to enter fully into the application of the system to every branch of science. The reader having been shown how to remember, with ease and certainty, words, ideas, figures and dates, and how to proceed with studies, etc., will find it highly advantageous to exercise his own ingenuity in applying the method to any particular study or subject that he may wish to master.

www.ingramcontent.com/pod-product-compliance
Lightning Source LLC
Chambersburg PA
CBHW022146090426
42742CB00010B/1408